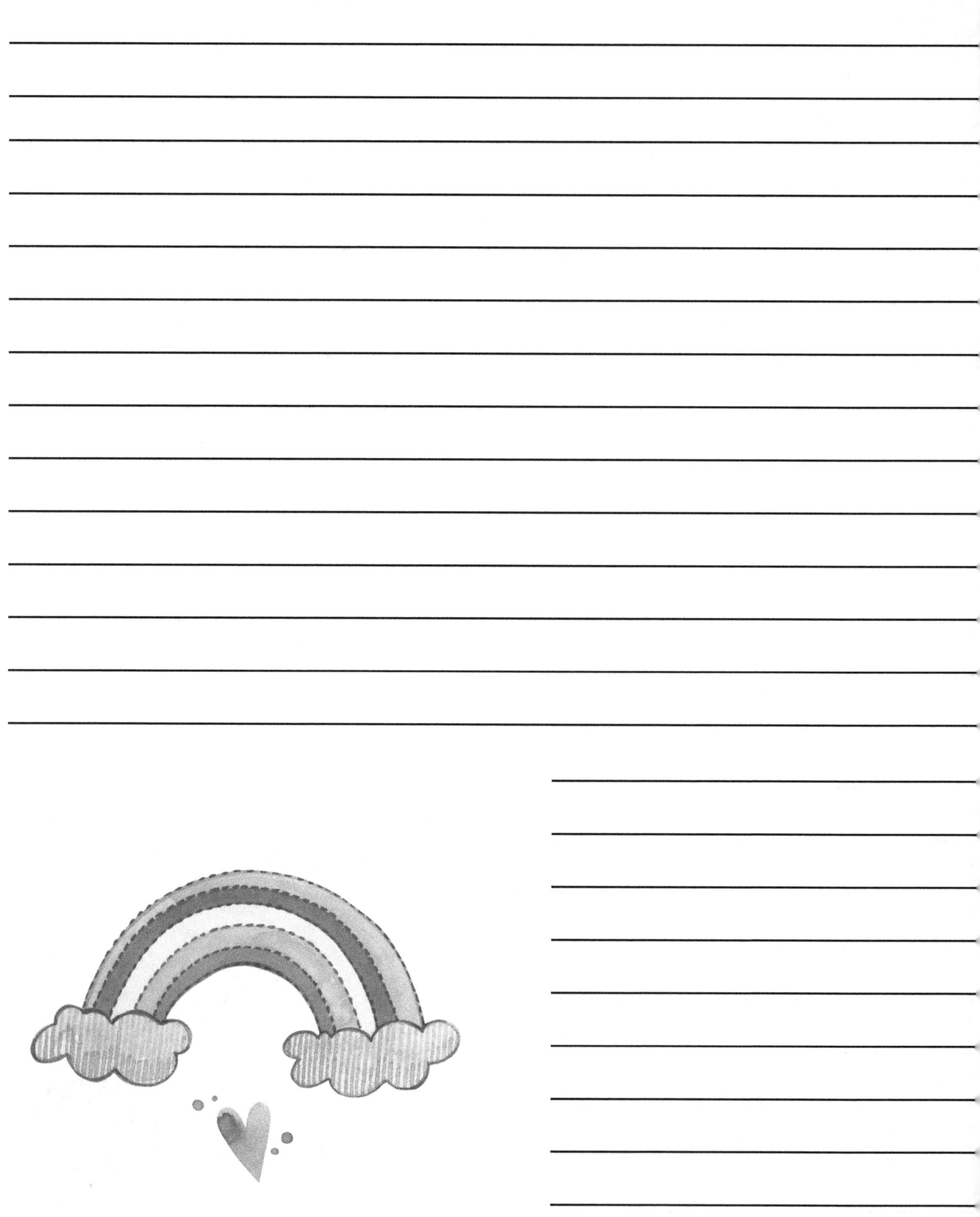

you + me = love

i love you

love you

LO
VE

you are loved

love
is all
you
need

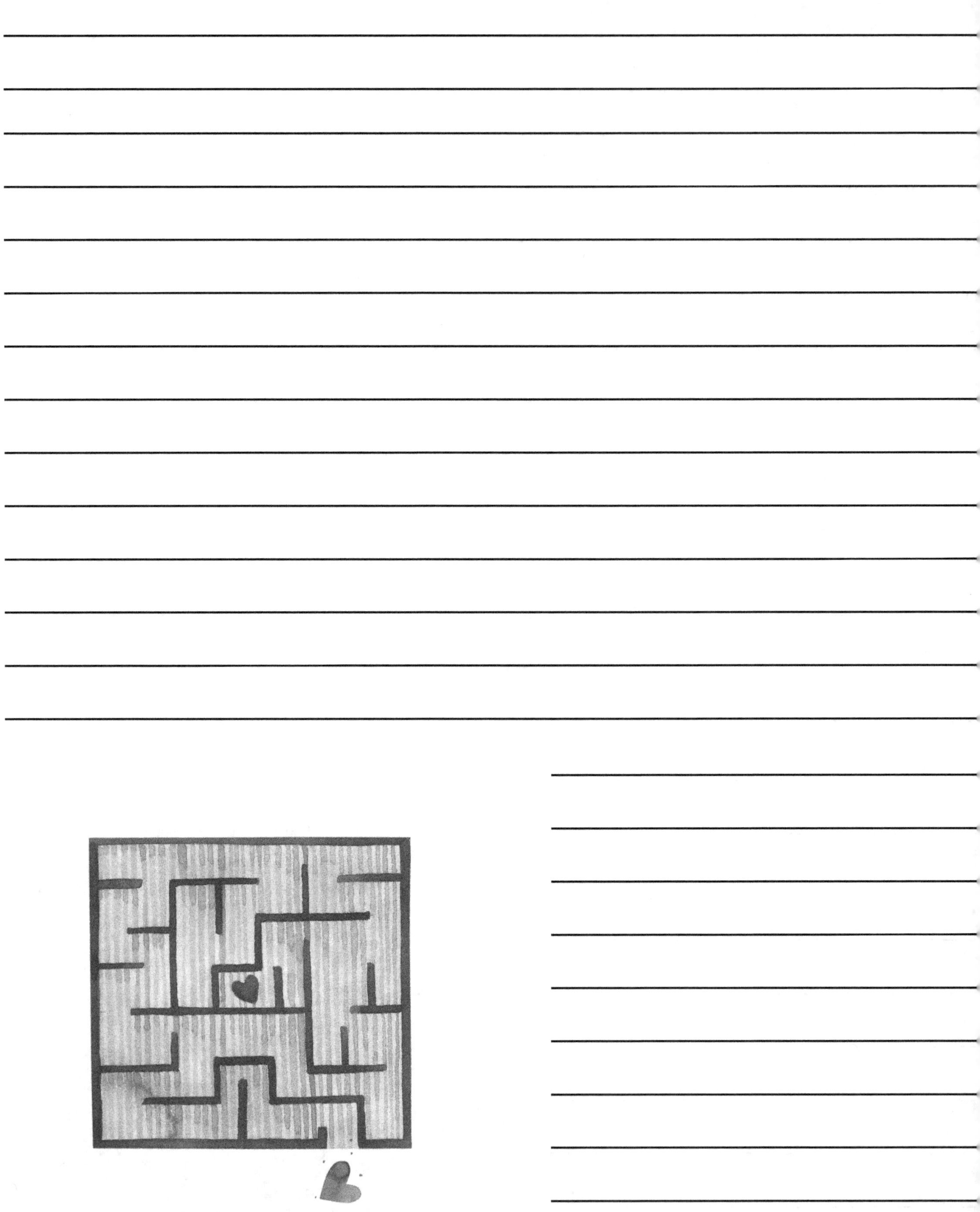

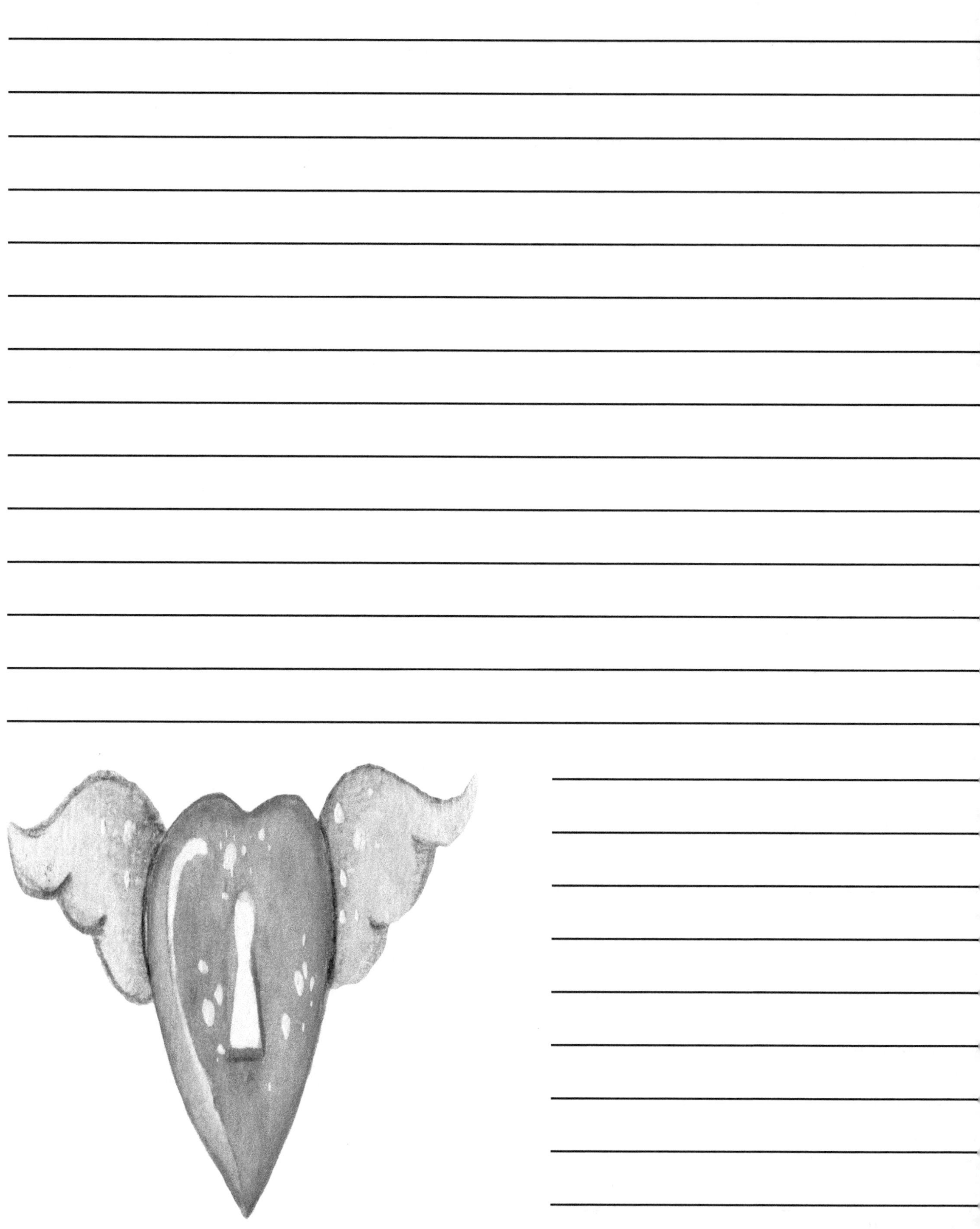

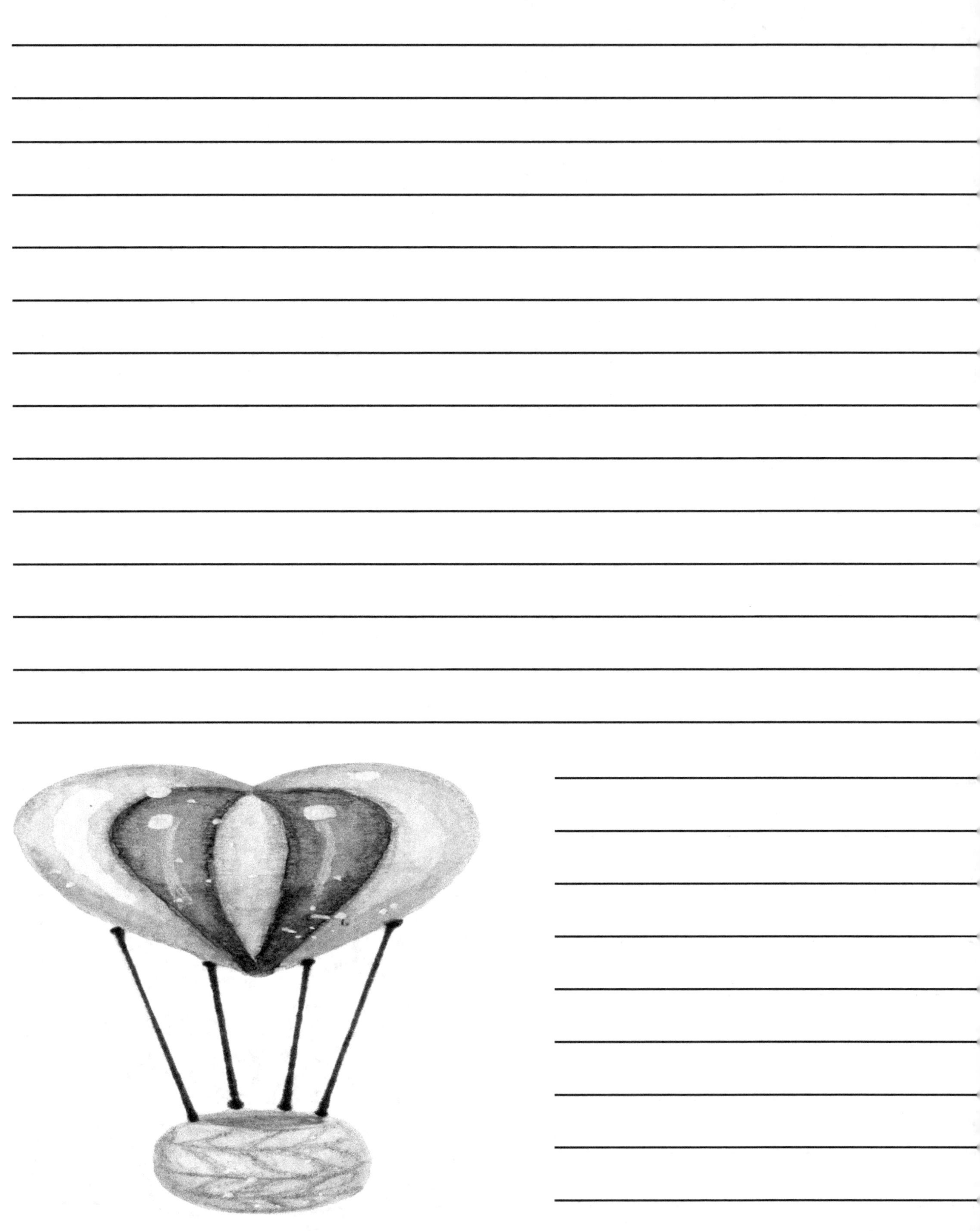

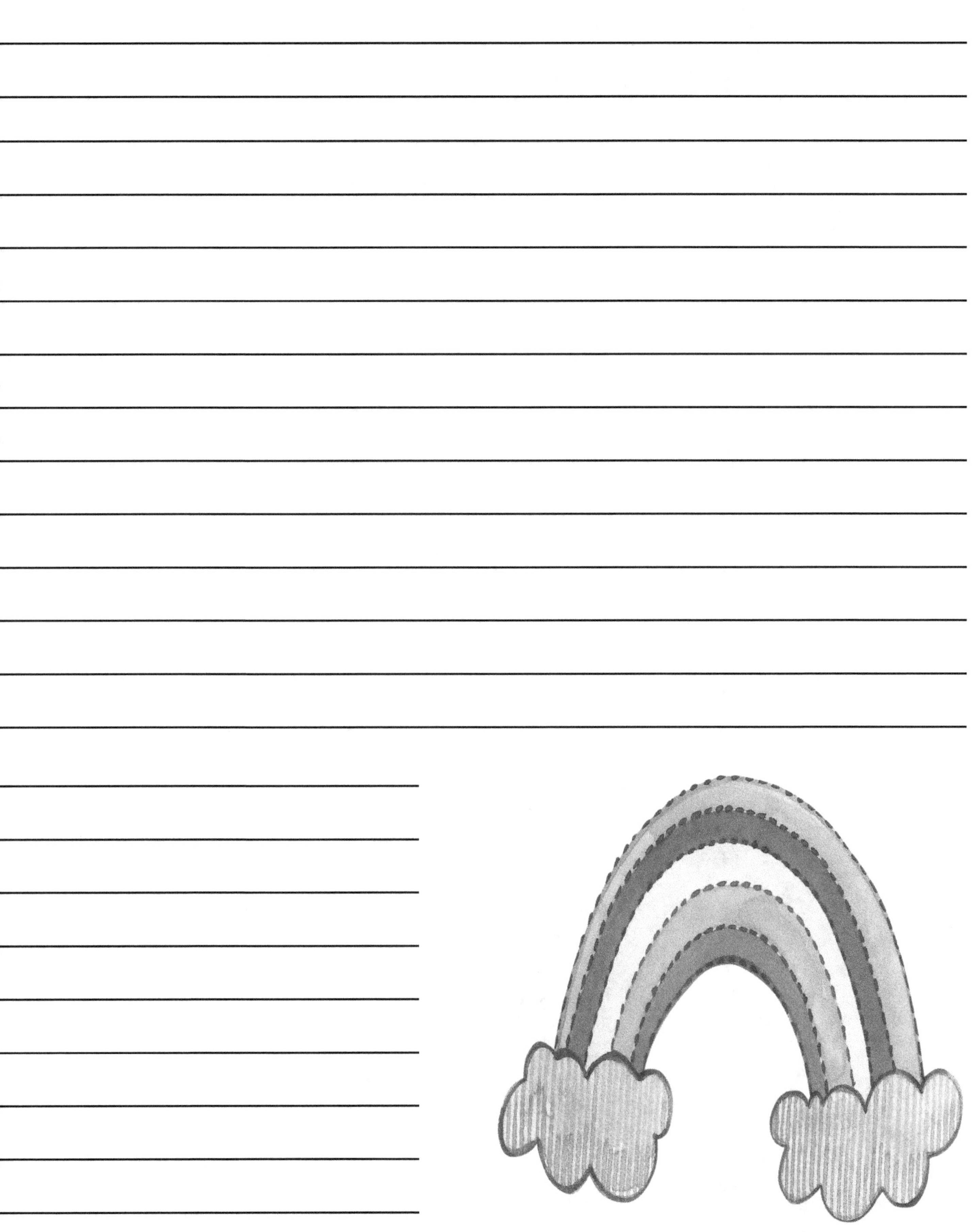

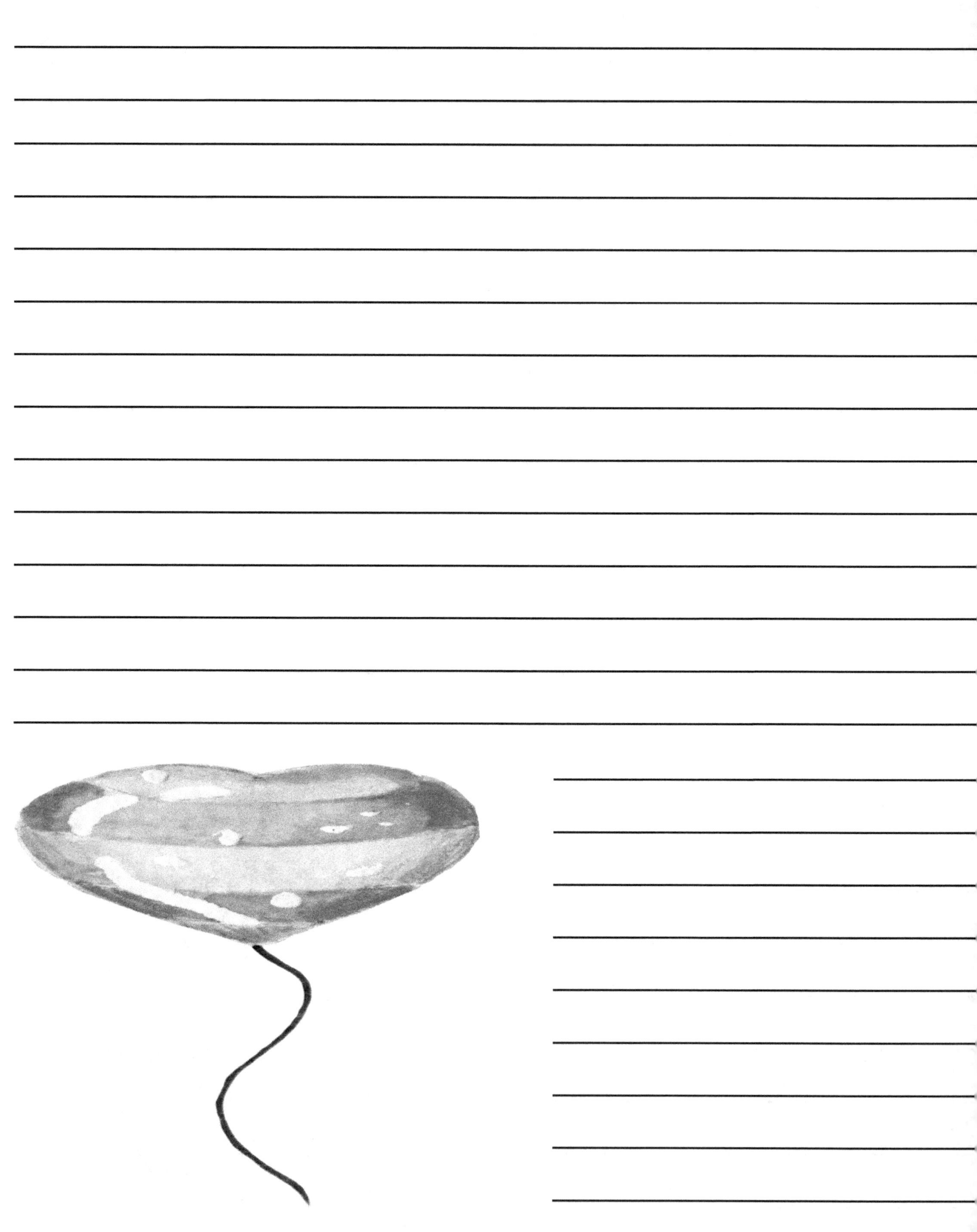

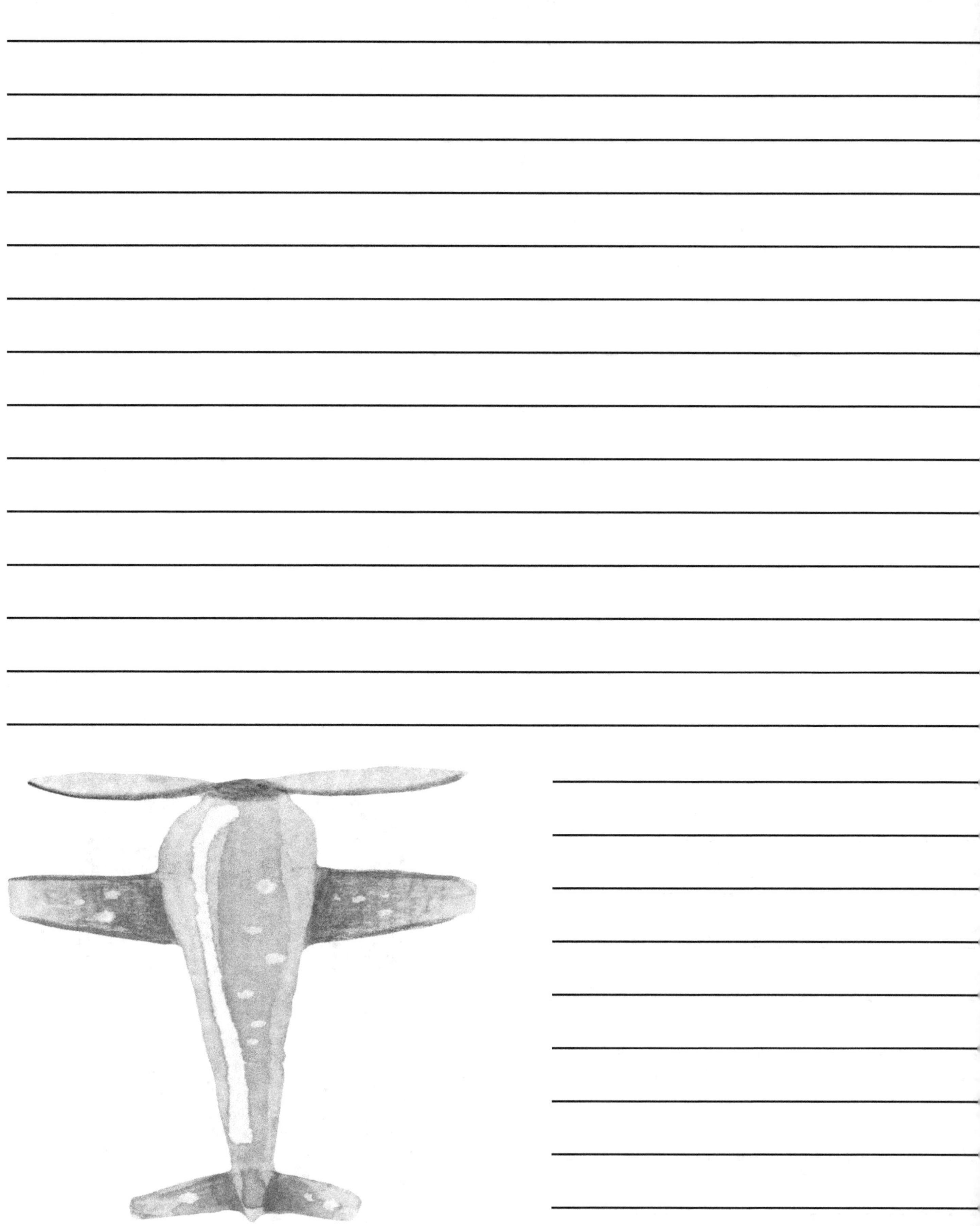

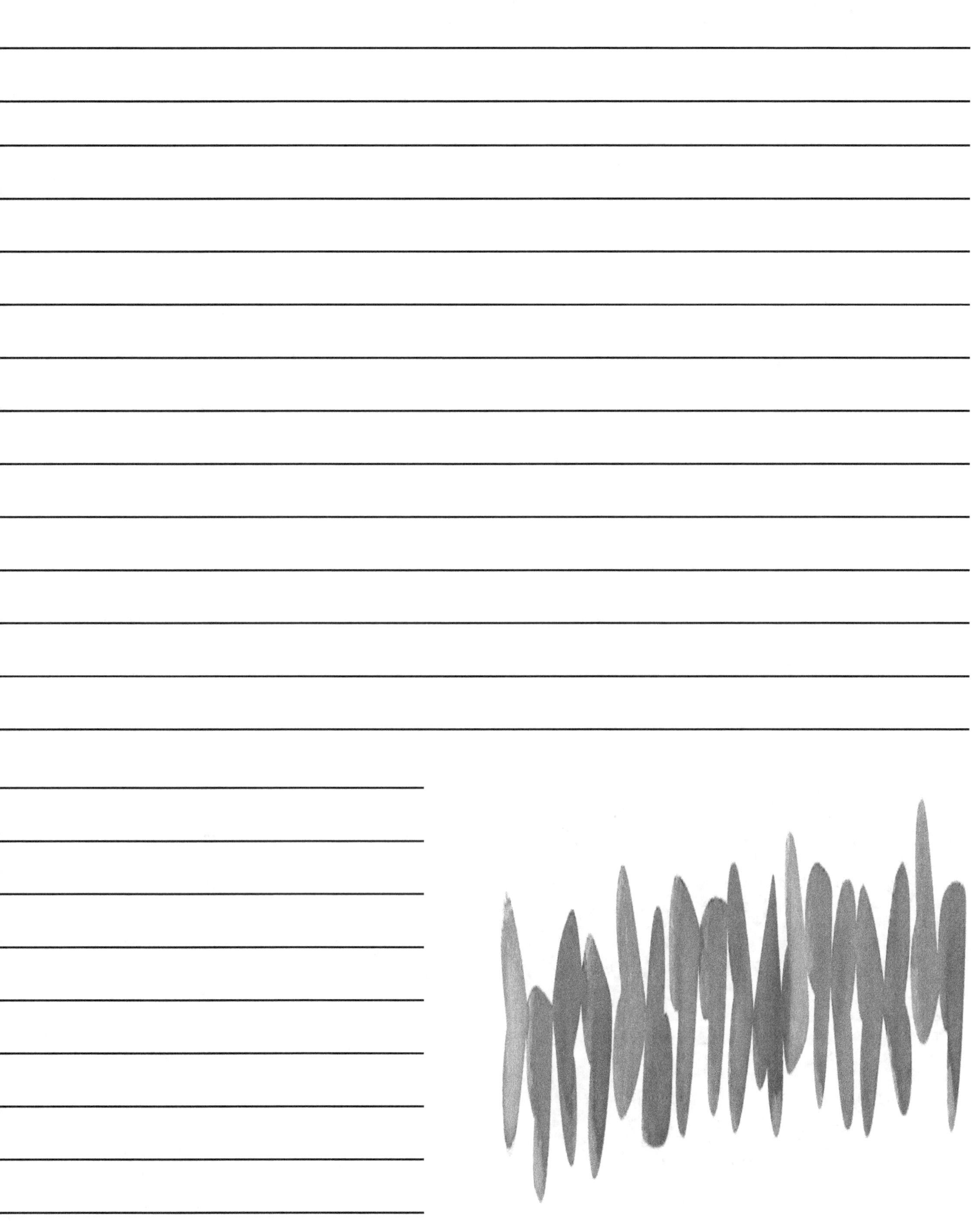

YAY

you
+
me
=
love

believe
IN
Love

i love
you

love
yourself

LO
VE

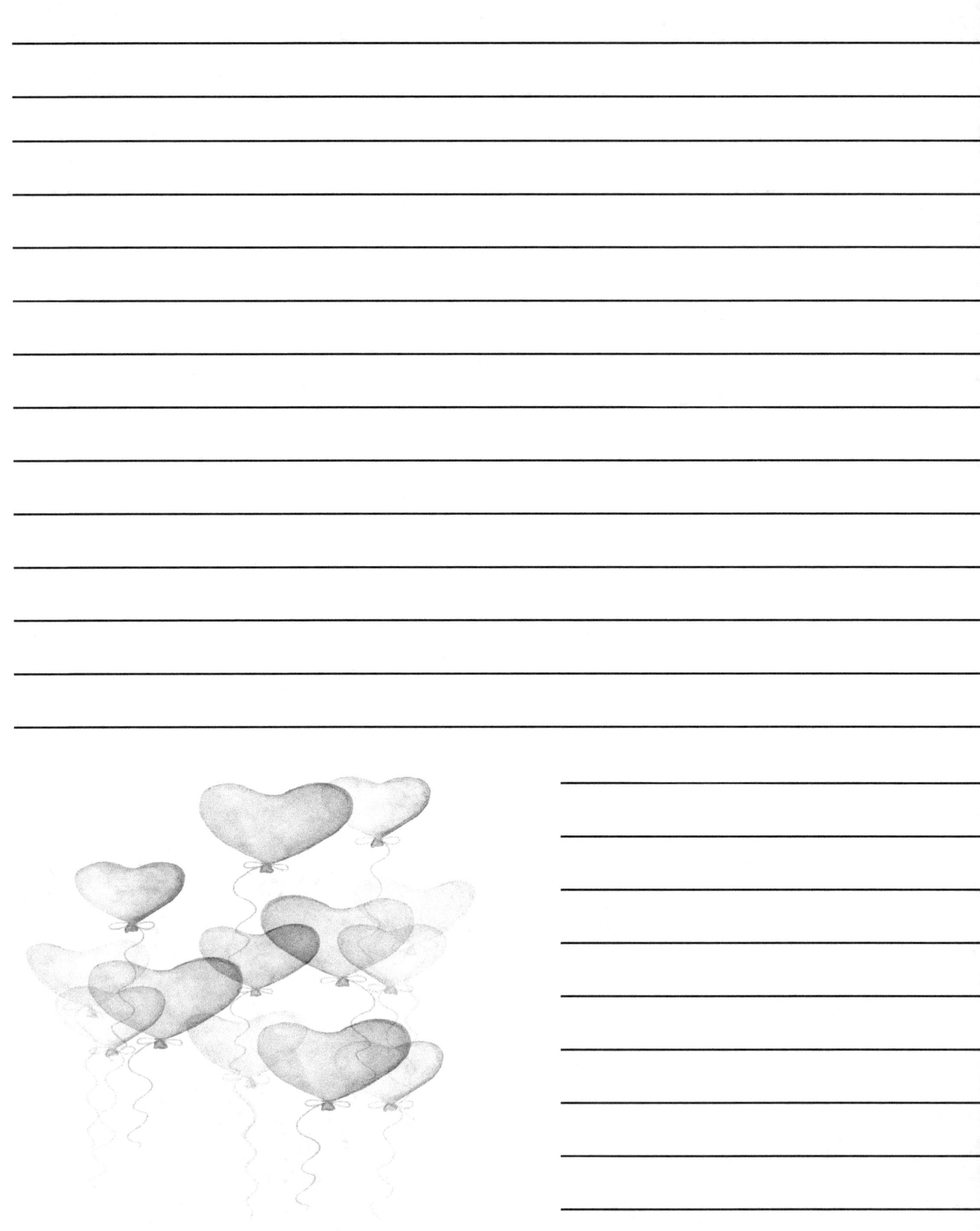

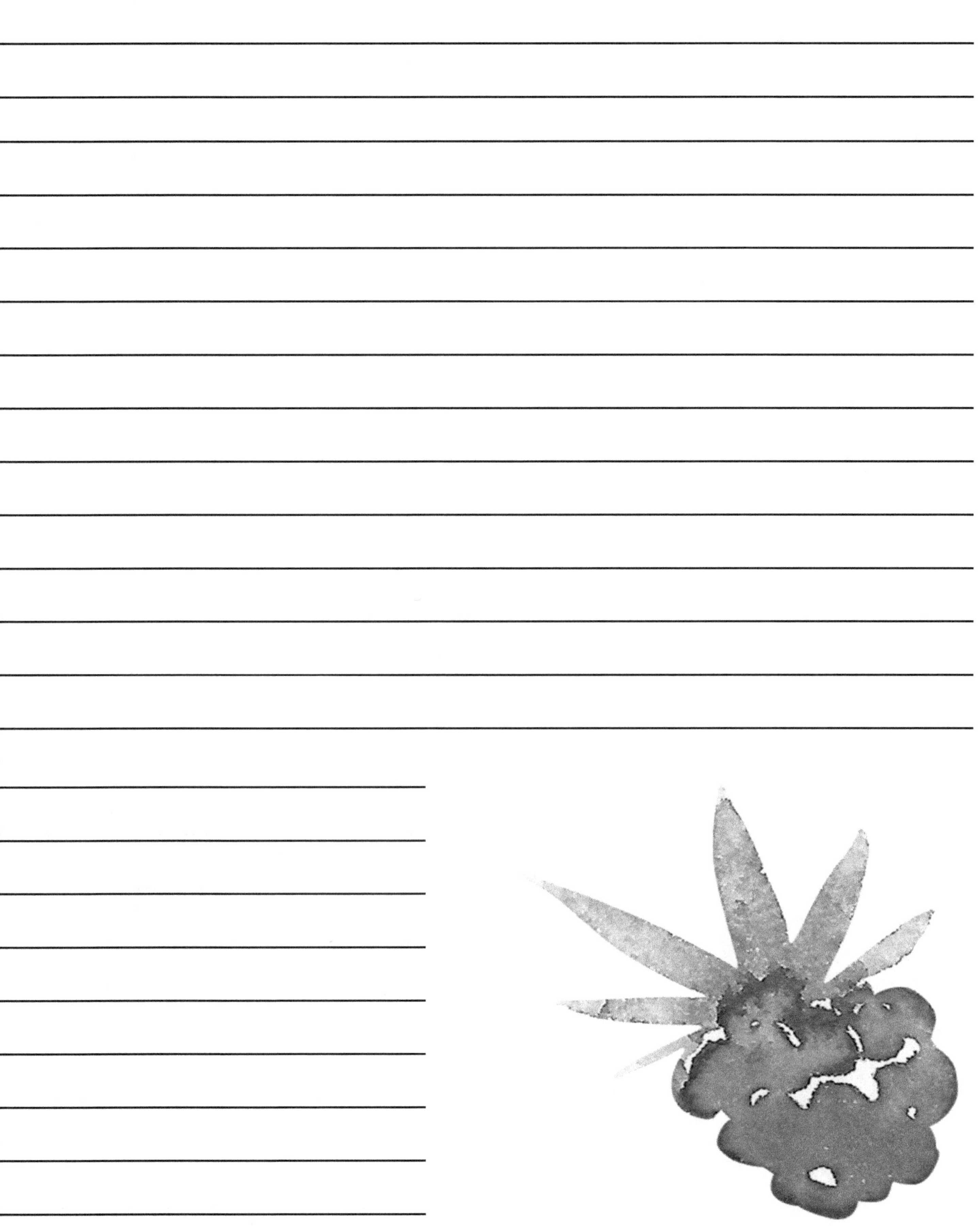

you are loved

love
is all
you
need

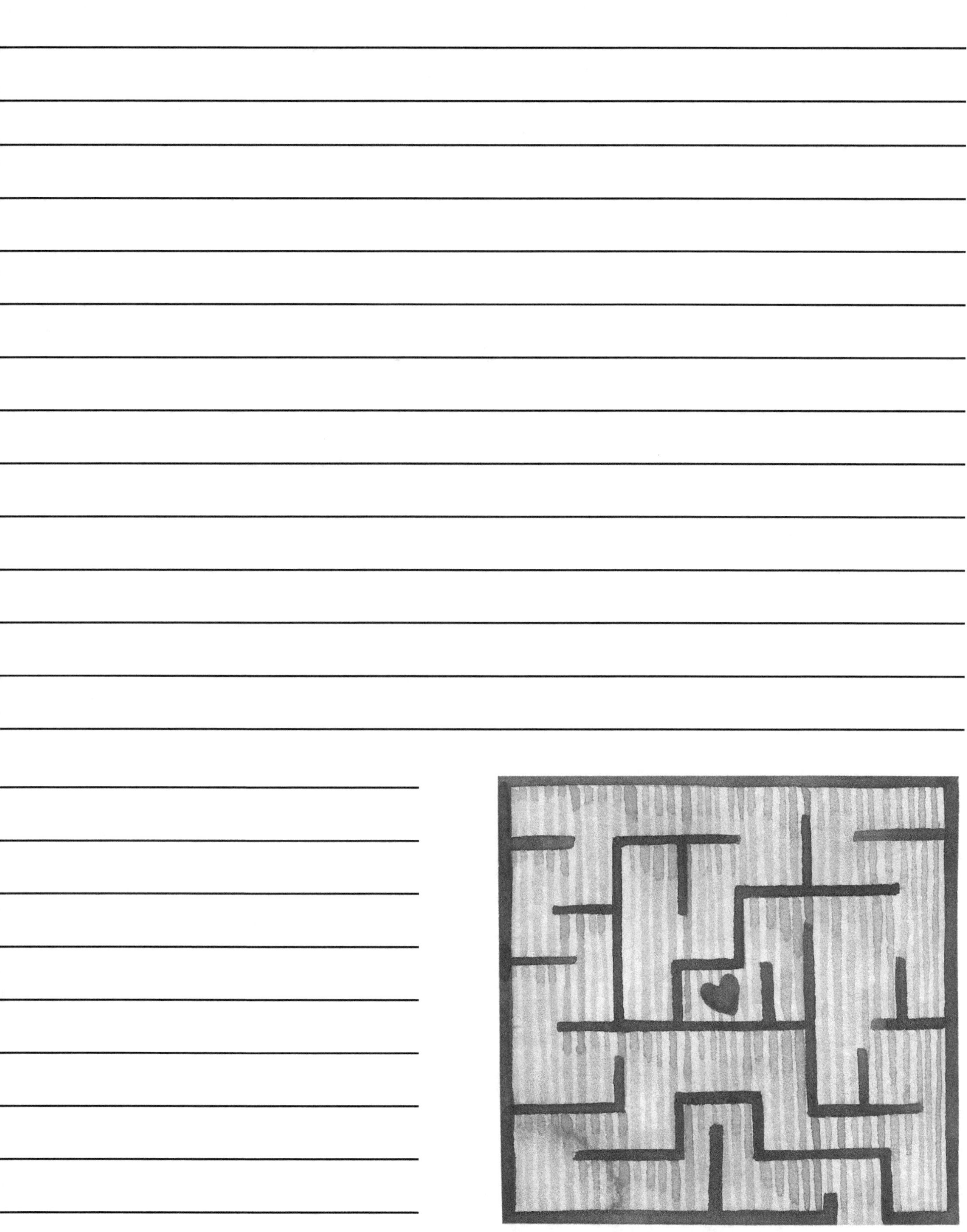

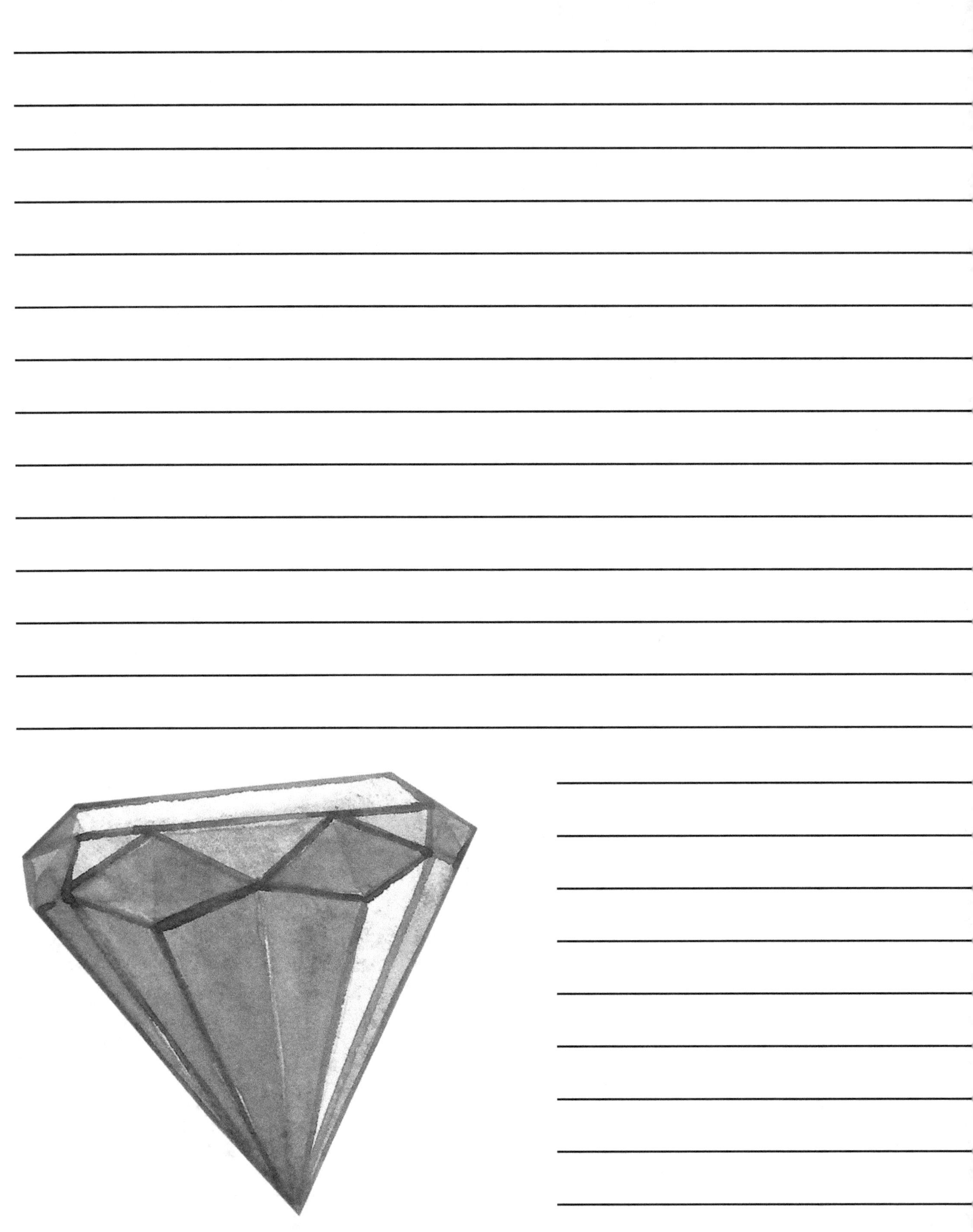

# - Acts of Kindness Tracker -

# - Acts of Kindness Tracker -

# *- Acts of Kindness Tracker -*

# Notes

# Notes

# Notes

www.ingramcontent.com/pod-product-compliance
Lightning Source LLC
LaVergne TN
LVHW080559200726
843510LV00004B/952